RISING STARS
Mathematics

Year
2

Practice Book A

Author: Paul Broadbent

ISBN: 978-1-78339-813-3
Text, design and layout © Hodder & Stoughton Ltd. (for its Rising Stars imprint) 2019

First published in 2016 by
Rising Stars, part of Hodder & Stoughton Ltd.
Reprinted 2019
An Hachette UK Company
Carmelite House
50 Victoria Embankment
London EC4Y 0DZ
www.risingstars-uk.com

Author: Paul Broadbent
Programme consultants: Cherri Moseley, Caroline Clissold, Paul Broadbent
Publishers: Fiona Lazenby and Alexandra Riley
Editorial: Jan Fisher, Aidan Gill
Project manager: Sue Walton

Series and character design: Steve Evans
Text design: Words & Pictures
Illustrations by Steve Evans

Cover design: Steve Evans and Words & Pictures

Printed by Ashford Colour Press Ltd
A catalogue record for this title is available from the British Library.

Contents

Comparing and ordering

1a Comparing numbers

YOU WILL NEED:
- number rods or interlocking cubes

Make these numbers with rods or cubes.

Tick the larger number in each pair.

30   50   60   80

40   90   90   10

70   60   80   50

50   20   20   70

2 Write these price labels in order. Start with the lowest amount.

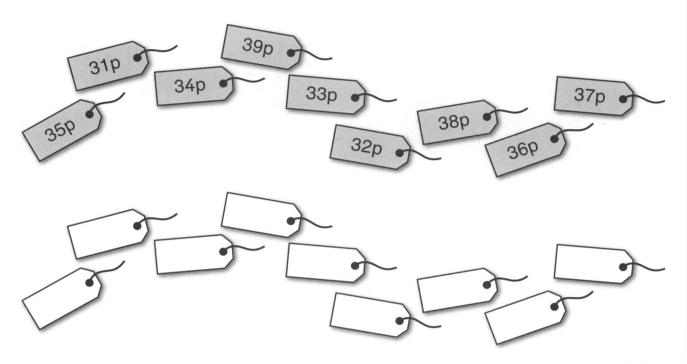

3 Write the numbers in the correct boxes.

a 67   76

⬚ is greater than ⬚

b 59   95

⬚ is greater than ⬚

c 43   34

⬚ is greater than ⬚

d 79   97

⬚ is less than ⬚

e 52   25

⬚ is less than ⬚

f 16   61

⬚ is less than ⬚

4 Put these numbers in order.

Start with the **greatest** number.

Use the 100 square to help you.

1	2	3	4	5	6	7	8	9	10
11	12	13	14	15	16	17	18	19	20
21	22	23	24	25	26	27	28	29	30
31	32	33	34	35	36	37	38	39	40
41	42	43	44	45	46	47	48	49	50
51	52	53	54	55	56	57	58	59	60
61	62	63	64	65	66	67	68	69	70
71	72	73	74	75	76	77	78	79	80
81	82	83	84	85	86	87	88	89	90
91	92	93	94	95	96	97	98	99	100

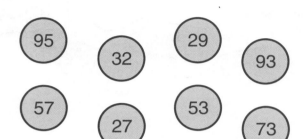

5 With 2 counters you can make 3 different numbers on this abacus.

> **YOU WILL NEED:**
> • **counters**

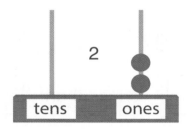 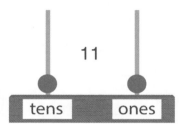 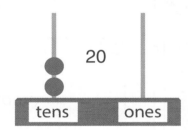

Use 3 counters to make different numbers on this abacus. Write them on the table.

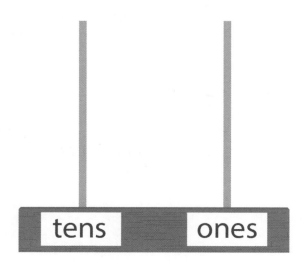

Numbers with 3 counters	Numbers with 4 counters	Numbers with 5 counters
3		
12		

Now try with 4 counters and then 5 counters.

1 Write the numbers shown.

a $\boxed{1\;0} \rhd \boxed{2} \rhd \boxed{}$ d $\boxed{4\;0} \rhd \boxed{6} \rhd \boxed{}$

b $\boxed{2\;0} \rhd \boxed{2} \rhd \boxed{}$ e $\boxed{4\;0} \rhd \boxed{9} \rhd \boxed{}$

c $\boxed{2\;0} \rhd \boxed{6} \rhd \boxed{}$ f $\boxed{8\;0} \rhd \boxed{9} \rhd \boxed{}$

2 Partition these into tens and ones.

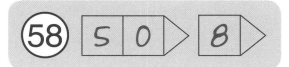

a (25) $\boxed{} \rhd \boxed{}$ d (46) $\boxed{} \rhd \boxed{}$

b (35) $\boxed{} \rhd \boxed{}$ e (59) $\boxed{} \rhd \boxed{}$

c (38) $\boxed{} \rhd \boxed{}$ f (79) $\boxed{} \rhd \boxed{}$

Partition each of these in 5 different ways.

51	54	57	58
50 and 1	50 and ☐	☐ and 7	☐ and 8
40 and 11	40 and ☐	☐ and 17	☐ and 18
30 and ☐	30 and ☐	☐ and 27	☐ and 28
20 and ☐	20 and ☐	☐ and 37	☐ and 38
10 and ☐	10 and ☐	☐ and 47	☐ and 48

4 Choose 3 different ways to partition these.

a

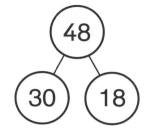
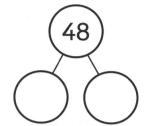
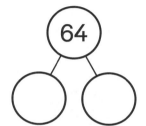

b

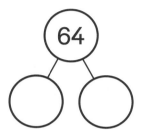

c

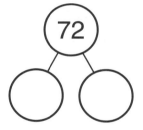
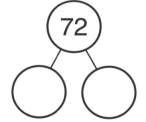

d

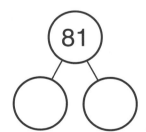
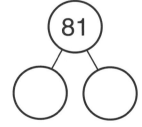

5 Use these numbers to make 56, 62, 76 and 32. You can only use each number once.

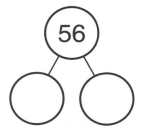 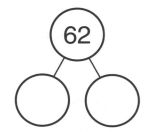

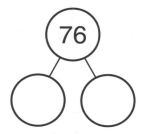 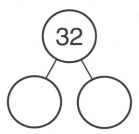

6 Complete these place value trees to help you solve these problems.

a If I spend 16p, how much do I have left?

b If I spend 40p, how much do I have left?

c If I spend 30p, how much do I have left?

d If I cut off 47 cm, how much wood is left?

|←—— 57 cm ——→|

e If I cut off 20 cm, how much string is left?

|←————— 95 cm —————→|

f If I cut off 38 cm, how much ribbon is left?

|←———— 88 cm ————→|

1 Partition these numbers into hundreds, tens and ones.

a (132)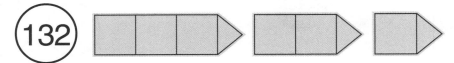

b (232)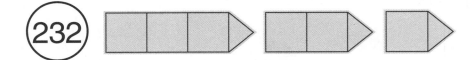

c (242)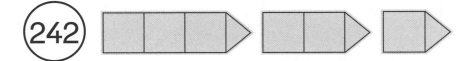

d (243)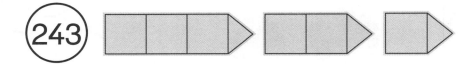

2 Look at the length and height of these.

128 cm   154 cm   215 cm   230 cm

Complete this using metres and centimetres.

	height 128 cm	a	___ metres	___ centimetres
	length 154 cm	b	___ metres	___ centimetres
	length 215 cm	c	___ metres	___ centimetres
	height 230 cm	d	___ metres	___ centimetres

3 Join the lengths that are the same.

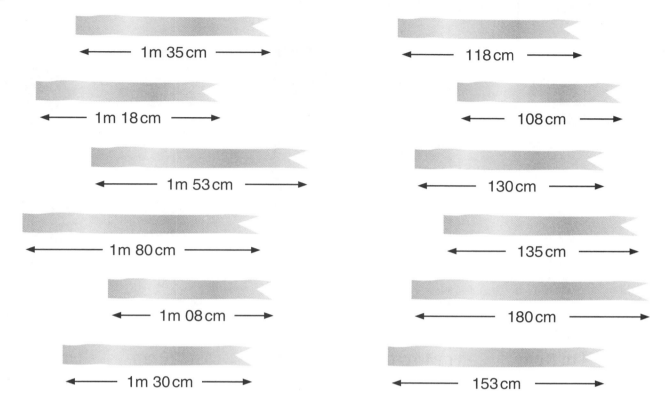

← 1m 35 cm →   ← 118 cm →

← 1m 18 cm →   ← 108 cm →

← 1m 53 cm →   ← 130 cm →

← 1m 80 cm →   ← 135 cm →

← 1m 08 cm →   ← 180 cm →

← 1m 30 cm →   ← 153 cm →

4 Colour each metre stick to show the fractions of 1 m.

a

$\frac{1}{4}$ m

b

$\frac{1}{2}$ m

c

$\frac{3}{4}$ m

These are the heights of 5 children. Write them in centimetres in order of height. Start with the tallest.

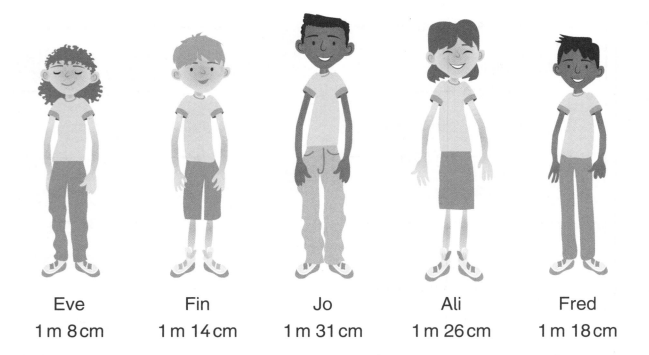

Eve	Fin	Jo	Ali	Fred
1 m 8 cm	1 m 14 cm	1 m 31 cm	1 m 26 cm	1 m 18 cm

	_____ cm
	_____ cm
	_____ cm
	_____ cm
	_____ cm

⭐ **1** Write the times shown on these clocks.

a

[o'clock]

c

[o'clock]

e

[o'clock]

b

[o'clock]

d

[o'clock]

f

[o'clock]

⭐ **2** Write the times shown on these clocks.

a

[half past]

c

[half past]

e

[half past]

b

[half past]

d

[half past]

f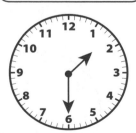

[half past]

3 How many hours are there between these times?

a

b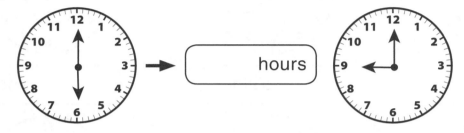

c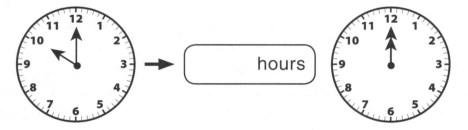

d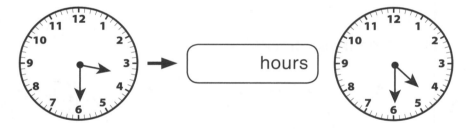

e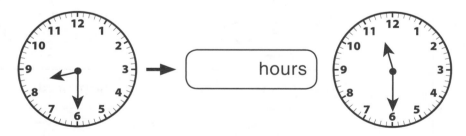

f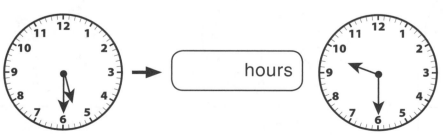

4 Order these from shortest time to longest time. Join them to the line to show the order.

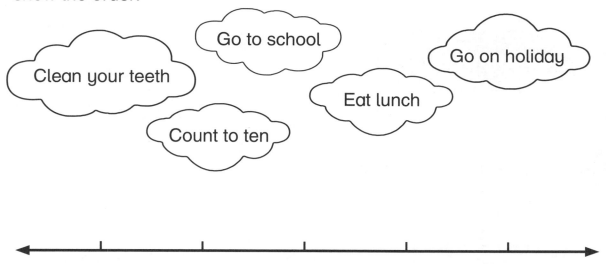

Shortest time   Longest time

5 How long is spent in total on each activity in each week?

	Monday	**Tuesday**	**Wednesday**	**Thursday**	**Friday**
1 hour	Badminton	Badminton	Football	Football	Badminton
1 hour	Keep fit	Keep fit	Keep fit	Keep fit	Keep fit
½ an hour	Yoga	Yoga	Yoga	Yoga	Yoga
½ an hour	Tennis	Basketball	Tennis	Basketball	Tennis
1 hour	Bowls		Bowls		

Activity	**Time (hours)**
Badminton	
Keep fit	
Yoga	
Tennis	
Bowls	
Basketball	
Football	

Addition and subtraction

2a Fact families

1 Write all the pairs that make 10.

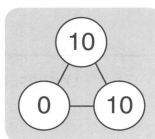

10
0 — 10

b

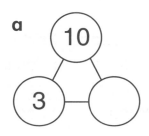

10
1

d
10
8

a
10
3

c
10
6

e
10
5

2 Complete these.

a 3 + 7 = 10
13 + 7 = ☐

c 6 + 4 = 10
16 + 4 = ☐

e 9 + 1 = 10
9 + 11 = ☐

b 5 + 5 = 10
5 + 15 = ☐

d 2 + 8 = 10
2 + 18 = ☐

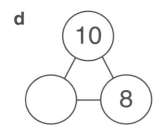

What do you notice?

3 Join the pairs that total 20.

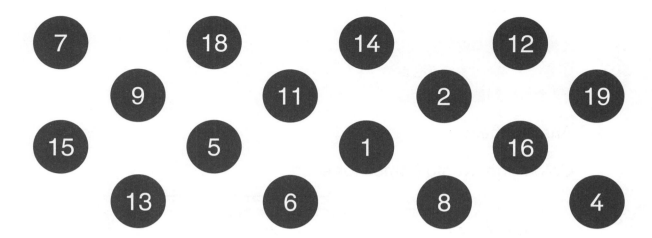

4 Use the bars to help complete each of these.

a

20	
4	16

4 + ( ) = ( )

( ) + 4 = ( )

( ) − ( ) = 4

( ) − 4 = ( )

b

20	
3	17

3 + ( ) = ( )

( ) + 3 = ( )

( ) − ( ) = 3

( ) − 3 = ( )

YOU WILL NEED:
- **3 coloured pencils**

Answer these.

Colour the twenty frames for each.

a

$4 + 8 = \boxed{}$

$\textcircled{4} + \textcircled{6} + 2 = \boxed{}$

b

$5 + 9 = \boxed{}$

$\textcircled{5} + \textcircled{5} + 4 = \boxed{}$

c

$7 + 6 = \boxed{}$

$\textcircled{7} + \textcircled{3} + 3 = \boxed{}$

d

$6 + 9 = \boxed{}$

$\textcircled{6} + \textcircled{4} + 5 = \boxed{}$

e

$7 + 4 = \boxed{}$

$\textcircled{7} + \textcircled{3} + 1 = \boxed{}$

2

Add these by counting on.
Use Base 10 apparatus to model each one.

a ⬚ = 13 + 6

| | | | | | | | | | |
| 10 | 11 | 12 | 13 | 14 | 15 | 16 | 17 | 18 | 19 | 20 |

b ⬚ = 24 + 4

| 20 | 21 | 22 | 23 | 24 | 25 | 26 | 27 | 28 | 29 | 30 |

c ⬚ = 22 + 7

| 20 | 21 | 22 | 23 | 24 | 25 | 26 | 27 | 28 | 29 | 30 |

d ⬚ = 31 + 6

| 30 | 31 | 32 | 33 | 34 | 35 | 36 | 37 | 38 | 39 | 40 |

3

Subtract these by counting back.
Use Base 10 apparatus to model each one.

a ⬚ = 19 – 6

| 10 | 11 | 12 | 13 | 14 | 15 | 16 | 17 | 18 | 19 | 20 |

b ⬚ = 28 – 4

| 20 | 21 | 22 | 23 | 24 | 25 | 26 | 27 | 28 | 29 | 30 |

c ⬚ = 29 – 7

| 20 | 21 | 22 | 23 | 24 | 25 | 26 | 27 | 28 | 29 | 30 |

d ⬚ = 37 – 6

| 30 | 31 | 32 | 33 | 34 | 35 | 36 | 37 | 38 | 39 | 40 |

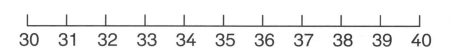

Use Base 10 apparatus to model each one.

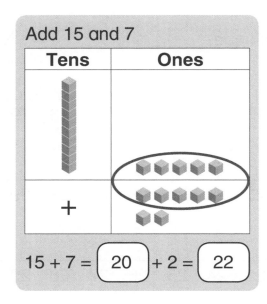

Add 15 and 7

Tens	Ones
+	

15 + 7 = [20] + 2 = [22]

a Add 16 + 8

Tens	Ones
+	

16 + 8 = 20 + ( ) = ( )

c Add 24 + 9

Tens	Ones
+	

24 + 9 = 30 + ( ) = ( )

b Add 25 + 6

Tens	Ones
+	

25 + 6 = 30 + ( ) = ( )

d Add 36 + 8

Tens	Ones
+	

36 + 8 = 40 + ( ) = ( )

5 Answer these.

Use Base 10 apparatus to model each one.

YOU WILL NEED:
• **Base 10 apparatus**

Subtract 8 from 24

Tens	Ones

Tens	Ones

Tens	Ones

$\boxed{24} - 8$

$\textcircled{10} + \textcircled{14} - 8$

$10 + \textcircled{6} = 16$

a   Subtract 7 from 25.

$25 - 7 = \boxed{}$

b   Subtract 6 from 32.

$32 - 6 = \boxed{}$

c   Subtract 8 from 36.

$36 - 8 = \boxed{}$

d   Subtract 5 from 32.

$32 - 5 = \boxed{}$

Tens	Ones

23

YOU WILL NEED:
- **interlocking cubes**
- **counters**
- **coins**

Answer these. Use cubes, counters and coins to help you.

a Sam has 15 cubes. Jo has 8 more cubes than Sam.
How many cubes does Jo have?

b Martha has 22 counters. Ravi has 6 fewer counters than Martha.
How many counters does Ravi have?

c Ben has 26 coins. Jasmine has 7 more coins than Ben.
How many coins does Jasmine have?

d Kate has 6 fewer cubes than Mark. Mark has 21 cubes.
How many cubes does Kate have?

e Matt and Ali have 16 counters altogether. Ali has got 9 counters.
How many counters does Matt have?

f Joe has 8 more coins than George. George has got 28 coins.
How many coins does Joe have?

1

YOU WILL NEED:
- **3 coloured pencils**

Answer these.

Look for pairs that total 10. Use three colours to show the numbers on the twenty frames.

a $5 + 6 + 4 = \boxed{}$

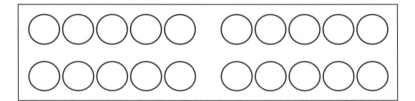

b $2 + 3 + 8 = \boxed{}$

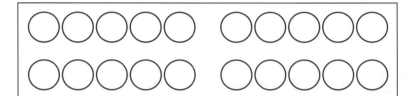

c $7 + 3 + 5 = \boxed{}$

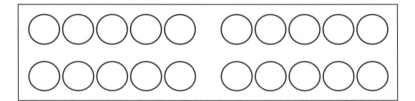

d $4 + 8 + 6 = \boxed{}$

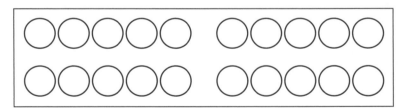

e $4 + 6 + 2 = \boxed{}$

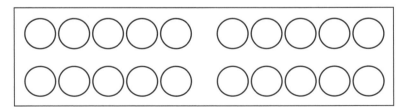

f $3 + 6 + 7 = \boxed{}$

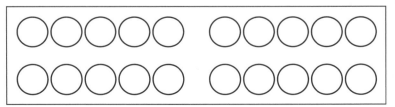

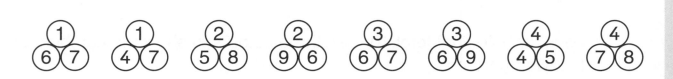

⭐ **2** Join each number to their total.

12 14 16 18

13 15 17 19

① ⑥⑦ ① ④⑦ ② ⑤⑧ ② ⑨⑥ ③ ⑥⑦ ③ ⑥⑨ ④ ④⑤ ④ ⑦⑧

⭐ **3** Answer these.

a $3 + \boxed{} + 5 = 5 + 7$

c $4 + 5 + \boxed{} = 6 + 9$

b $\boxed{} + 3 + 4 = 4 + 5$

d $5 + \boxed{} + 7 = 7 + 11$

Find 4 different ways to total 18 with 3 numbers.

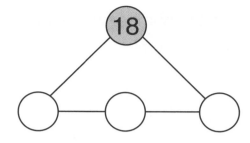

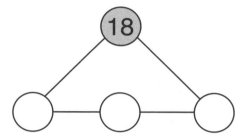

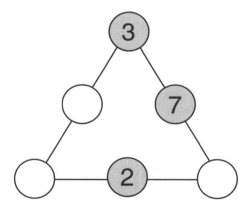

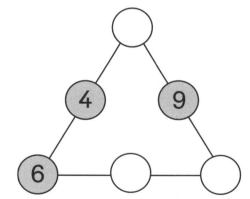

5 Complete so that the sum of the numbers on each line is 15.

a

b

1 The top number is the total of the two bottom numbers. Write the missing numbers.

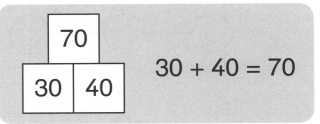

a

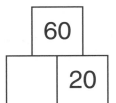

c

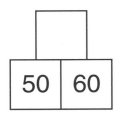

e

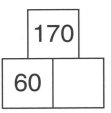

b

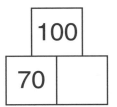

d

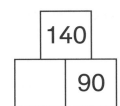

f

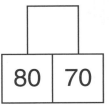

  2 Write the missing numbers.

a $90 - \boxed{} = 10$ **b** $100 - \boxed{} = 90$

$90 - \boxed{} = 30$ $120 - \boxed{} = 90$

$90 - \boxed{} = 50$ $140 - \boxed{} = 90$

$90 - \boxed{} = 70$ $160 - \boxed{} = 90$

3

Add these.

Use Base 10 apparatus to help you.

a   37 + 20 = ☐   d   57 + 50 = ☐

b   37 + 40 = ☐   e   59 + 50 = ☐

c   57 + 40 = ☐   f   59 + 70 = ☐

4   Answer these. Partition them first to subtract the tens.

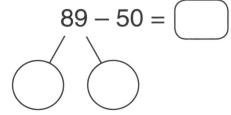

63 − 40 = 23
60   3

a   73 − 40 = ☐   d   68 − 30 = ☐

b   79 − 50 = ☐   e   88 − 50 = ☐

c   89 − 50 = ☐   f   99 − 60 = ☐

5 Write the new price for each item.

a

£87

£**40 off**
new price

£

c

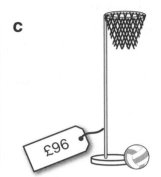

£96

£**30 off**
new price

£

b

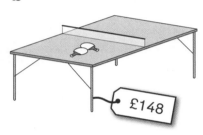

£148

£**70 off**
new price

£

d

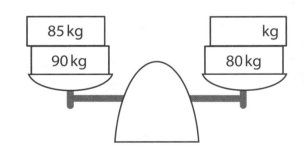

£131

£**60 off**
new price

£

6 Write the missing masses so these scales are balanced.

a

| 72 kg | | kg |
| 30 kg | | 52 kg |

b

| 85 kg | | kg |
| 90 kg | | 80 kg |

Shapes all around us

3a Patterns

1 Draw the next 3 shapes for each pattern.

a

b

c

d

2

YOU WILL NEED:
- **3 coloured pencils**

Colour these to make your own repeating pattern. Use 2 or 3 colours.

a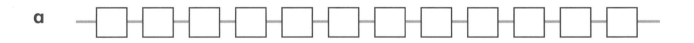

b

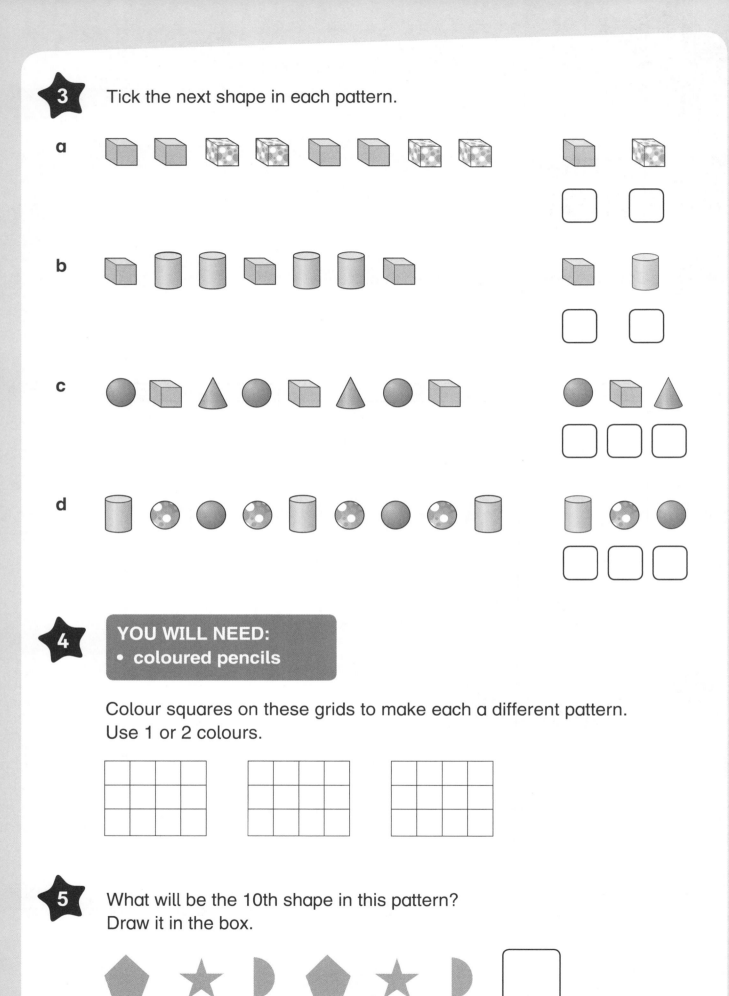

3 Tick the next shape in each pattern.

a

b

c

d

4 YOU WILL NEED:
- coloured pencils

Colour squares on these grids to make each a different pattern.
Use 1 or 2 colours.

5 What will be the 10th shape in this pattern?
Draw it in the box.

1 Cross out the odd shape in each set. Write the name of the set of shapes.

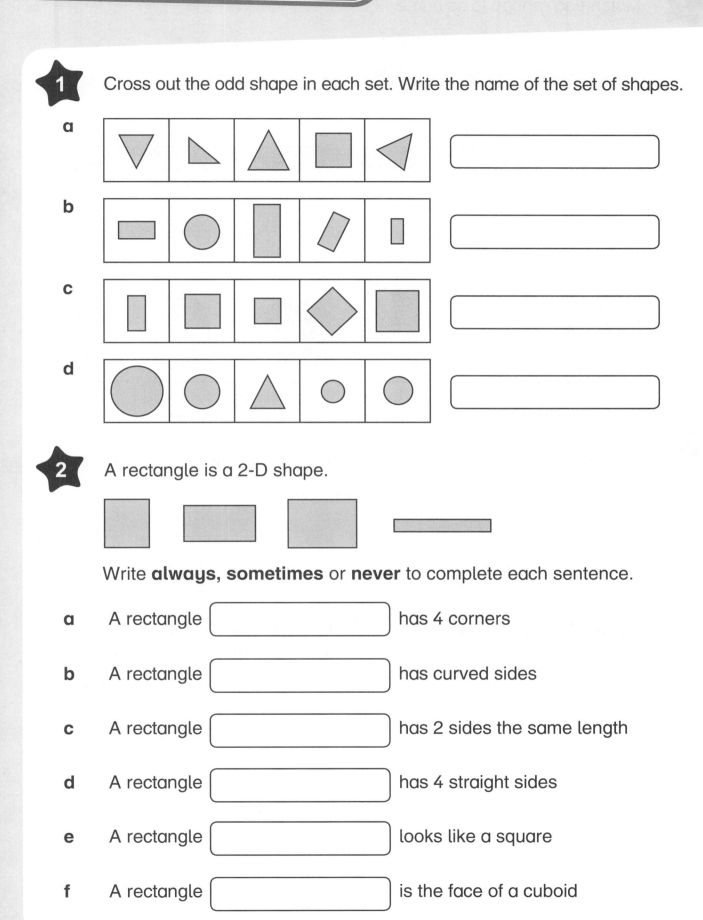

a

b

c

d

2 A rectangle is a 2-D shape.

Write **always, sometimes** or **never** to complete each sentence.

a   A rectangle [   ] has 4 corners

b   A rectangle [   ] has curved sides

c   A rectangle [   ] has 2 sides the same length

d   A rectangle [   ] has 4 straight sides

e   A rectangle [   ] looks like a square

f   A rectangle [   ] is the face of a cuboid

3 Match each shape to its name.

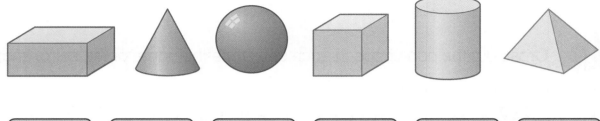

| cube | cuboid | cylinder | sphere | pyramid | cone |

4 Write the name of each shape.

| | | | | | |

5 Choose some different 3-D shapes that only have straight edges.

Look at them carefully and complete this table.

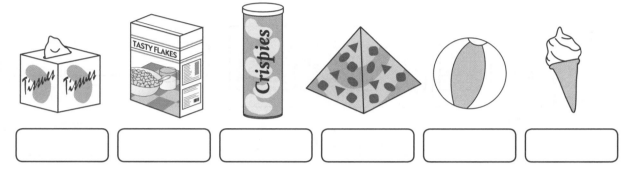

Name of shape	Shapes of faces (draw them)	Number of faces	Number of vertices	Number of edges

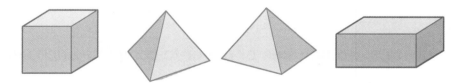

1 Look at these pictures. Tick the symmetrical shapes.

a

☐ ☐

d

☐ ☐

b

☐ ☐

e

☐ ☐

c

☐ ☐

f
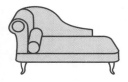
☐ ☐

2 Draw 1 line of symmetry on each shape.

3 Each drawing can be made symmetrical.

Complete each drawing using the line of symmetry.

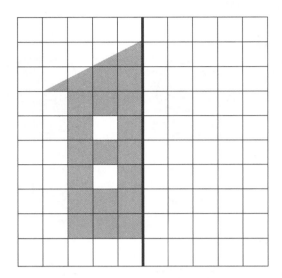

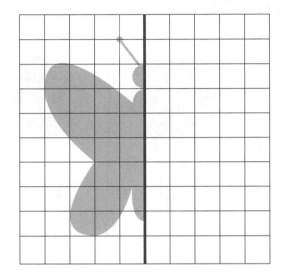

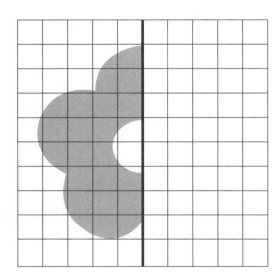

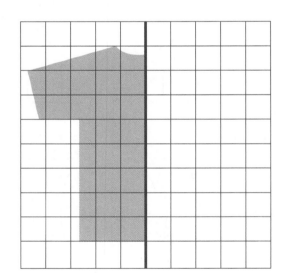

This pattern is symmetrical.

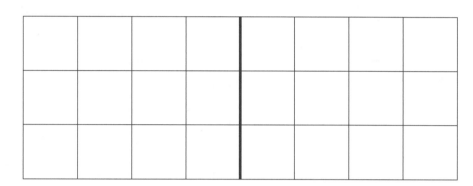

Draw your own symmetrical patterns on these grids.

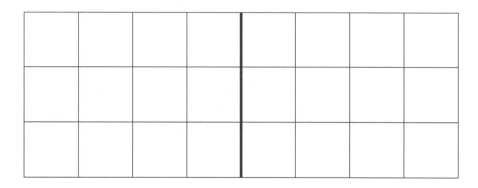

Number and measurement

4a Less than and greater than

 1 Write the missing numbers in each sequence.

a   17   18   [ ]   20   21   22   [ ]   [ ]   25

b   [ ]   54   55   56   57   [ ]   59   [ ]   61

c   88   89   90   [ ]   [ ]   93   [ ]   95   96

d   65   66   [ ]   68   [ ]   [ ]   71   72   73

e   31   [ ]   33   34   35   36   [ ]   38   [ ]

 2 Write < or > in each of these.

Remember: < means 'is less than' and > means 'is greater than'.

a   38 [ ] 51   d   25 [ ] 75   g   23 [ ] 50

b   62 [ ] 60   e   84 [ ] 48   h   92 [ ] 87

c   19 [ ] 41   f   56 [ ] 93   i   47 [ ] 16

3 Choose any number from the box to complete each of these.

14
35
98
22
72
50

a [ ] < 53

e [ ] < 23

b 85 > [ ]

f 49 < [ ]

c [ ] > 37

g [ ] < 74

d 61 > [ ]

h 93 < [ ]

4 Complete these by writing <, > or = to compare lengths.

a 28 cm [ ] 2 m

e 50 cm [ ] $\frac{1}{2}$ m

b 100 cm [ ] 1 m

f 140 cm [ ] 1 m 4 cm

c 17 cm [ ] 170 cm

g 1 m 80 cm [ ] 180 cm

d 1 $\frac{1}{2}$ m [ ] 15 cm

h 9 cm [ ] 9 m

5 Complete these by writing the correct symbol <, > or =.

a $5 + 6$ ⬚ 12

b 9 ⬚ $3 + 4$

c 11 ⬚ $4 + 7$

d $9 - 5$ ⬚ 3

e $8 - 2$ ⬚ 10

f 7 ⬚ $5 - 2$

 6 Write different possible missing numbers for these.

a $15 -$ ⬚ > 8

$15 -$ ⬚ > 8

$15 -$ ⬚ > 8

$15 -$ ⬚ > 8

b ⬚ $+ 12 < 20$

⬚ $+ 12 < 20$

⬚ $+ 12 < 20$

⬚ $+ 12 < 20$

4b **How much?**

⭐**1** Look at these balances. Answer the questions.

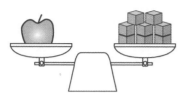 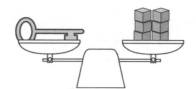 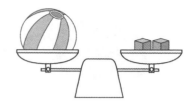

a Which is heavier, the apple or the key?

b How many balls would weigh about the same as a key? [ ] balls

c What is the mass of 2 keys? [ ] cubes

d How many cubes would balance 3 apples? [ ] cubes

e Which is heavier, 5 balls or 2 apples?

f What is the mass of an apple and a key? [ ] cubes

⭐**2** How much does each parcel weigh?

a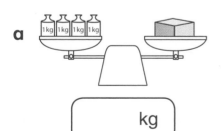

[ ] kg

c

[ ] kg

e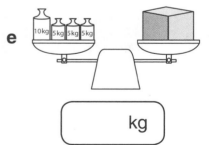

[ ] kg

b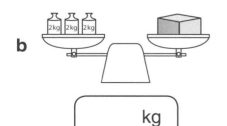

[ ] kg

d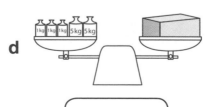

[ ] kg

f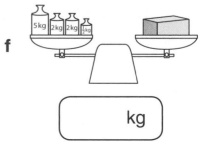

[ ] kg

3 Here are the masses of 5 rabbits.

Write them in order of weight, starting with the **heaviest**.

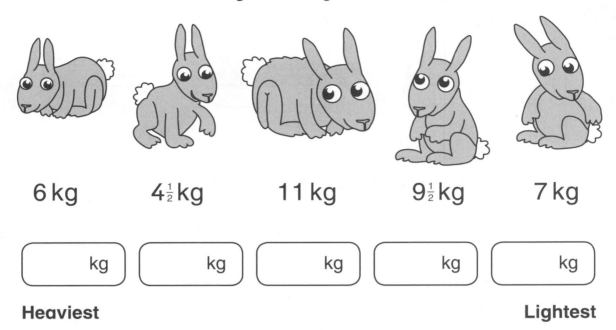

6 kg   4½ kg   11 kg   9½ kg   7 kg

| kg | kg | kg | kg | kg |

Heaviest   **Lightest**

4 Write the capacity of each of these in order.

Start with the greatest amount.

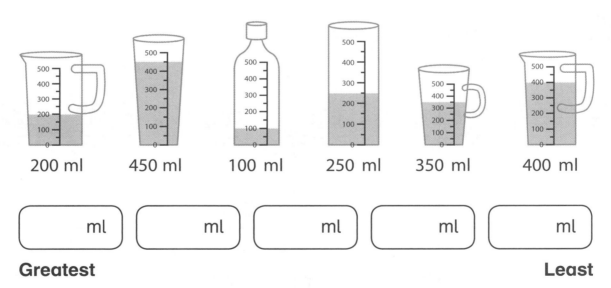

200 ml   450 ml   100 ml   250 ml   350 ml   400 ml

| ml | ml | ml | ml | ml |

Greatest   **Least**

5

YOU WILL NEED:
- litre jug
- different containers
- water

Use the litre jug to work out whether each container holds about the same as, more than or less than 1 litre of water.

Complete this chart.

About the same as 1 litre	Hold less than 1 litre	Hold more than 1 litre

6

YOU WILL NEED:
- litre jug
- different containers
- water

Use the containers that hold less than 1 litre from the chart above.

How many of each container are needed to fill the 1 litre jug?

1 Write the times shown on these clocks.

a

o'clock

c

o'clock

e

half past

b

o'clock

d

half past

f

half past

2 Draw hands on these clocks to show the times.

 One hour **after** 4 o'clock

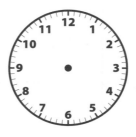 One hour **before** 11 o'clock

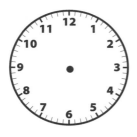 Half an hour **after** 7 o'clock

 Half an hour **before** half past 3

3 Write the times shown on these clocks.

a

quarter past

b

quarter past

c

quarter past

d

quarter past

e

quarter to

f

quarter to

g

quarter to

h

quarter to

4 Complete these.

a   1 week = [**days**]

d $\frac{1}{2}$ hour = [**minutes**]

b   1 day = [**hours**]

e $\frac{1}{4}$ hour = [**minutes**]

c   1 hour = [**minutes**]

f $\frac{3}{4}$ hour = [**minutes**]

5 Work out how many minutes there are between these times.

a

 → [**minutes**]

b

 → [**minutes**]

c

  → [**minutes**]

d

  → [**minutes**]

5a Patterns in calculations

 1 Write the missing numbers.

a 1 + [ ] = 10

3 + [ ] = 10

5 + [ ] = 10

7 + [ ] = 10

9 + [ ] = 10

b [ ] + 2 = 15

[ ] + 4 = 15

[ ] + 6 = 15

[ ] + 8 = 15

[ ] + 10 = 15

 2 Complete these.

a 6 + 9 = [ ] + 6

b 8 + [ ] = 4 + 8

c 7 + 5 = 5 + [ ]

d 12 + [ ] = 8 + 12

3 Use the first fact to help answer the others.

a $6 + 3 = 9$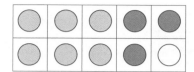

$16 + 3 = \boxed{}$ $36 + 3 = \boxed{}$ $66 + 3 = \boxed{}$

$6 + 13 = \boxed{}$ $6 + 33 = \boxed{}$ $6 + 63 = \boxed{}$

b $4 + 4 = 8$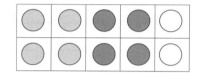

$14 + 4 = \boxed{}$ $24 + 4 = \boxed{}$ $54 + 4 = \boxed{}$

$4 + 14 = \boxed{}$ $4 + 24 = \boxed{}$ $4 + 54 = \boxed{}$

4 Group the cubes and rods to show your answers.

$3 + 7 = \boxed{10}$ $30 + 70 = \boxed{100}$

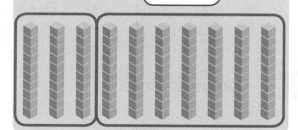

a $4 + 6 = \boxed{}$ $40 + 60 = \boxed{}$

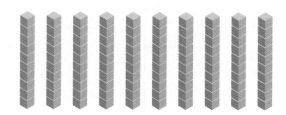

b $5 + \boxed{} = 10$ $50 + \boxed{} = 100$

   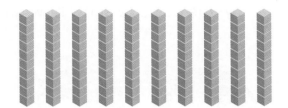

c $\boxed{} + 8 = 10$ $\boxed{} + 80 = 100$

   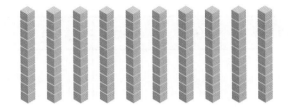

d $9 + \boxed{} = 10$ $90 + \boxed{} = 100$

   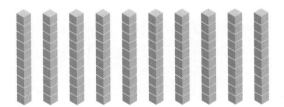

e $\boxed{} + 4 = 10$ $\boxed{} + 40 = 100$

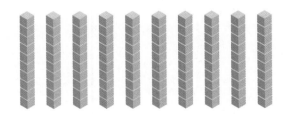

5 Make each total from 10p to 90p with any of these 4 coins.

Draw the coins to show how you make the amounts.

10p	20p	30p	40p	50p	60p	70p	80p	90p

6 Write the missing numbers. The missing numbers are the totals of the 2 outside circles.

a

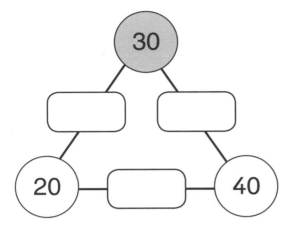

c

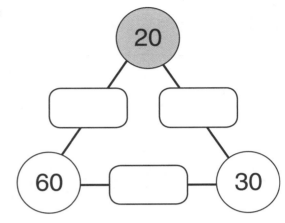

b

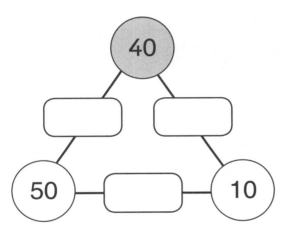

Try this one.

d

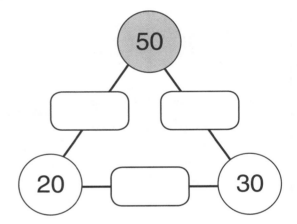

e

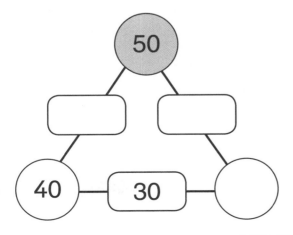

1 Write each total.

a [ ] p

b £[ ]

c [ ] p

d £[ ]

e [ ] p

f £[ ]

What do you notice?

2 Use any of these coins. Draw the coins to show these totals in 2 different ways.

a

56p

3 coins	5 coins

b

67p

4 coins	5 coins

c

73p

4 coins	6 coins

d

86p

5 coins	6 coins

3 Which silver coins could you use to pay for these?

Show 3 different ways.

a

55p

b

60p

c

85p

4

YOU WILL NEED:
- **four 10p coins**
- **five 20p coins**

Play 'Make 50p' with a partner.

- Take turns to place a 10p or 20p coin anywhere on the grid.
- If you make a line of 3 coins in any direction that total 50p, you score a point.
- Carry on until all the coins are used.
- Play again.
- The first player to 10 points is the winner.

1 Use this number line to help you find these totals.

30   35   40   45   50

a   34p + 8p = [p]   38p + 4p = [p]

b   £33 + £9 = [£   ]   £39 + £3 = [£   ]

c   39p + 7p = [p]   37p + 9p = [p]

d   £38 + £6 = [£   ]   £36 + £8 = [£   ]

2 Answer these.

a   2 ice lollies cost a total of 90p. The difference in price between them is 10p.

   What is the cost of each ice lolly?   [p]   [p]

b   2 coats cost a total of £30. One coat is £8 more than the other.

   What is the cost of each coat?   [£   ]   [£   ]

3 What is the difference in price between each of these?

Use the number lines to show your method.

a

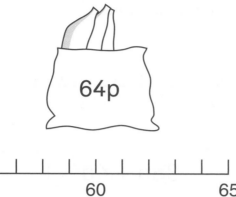

64p   59p

55   60   65

[   ] p

b

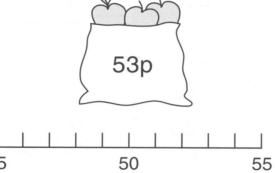

53p   48p

45   50   55

[   ] p

c

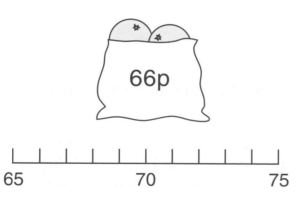

66p   72p

65   70   75

[   ] p

d

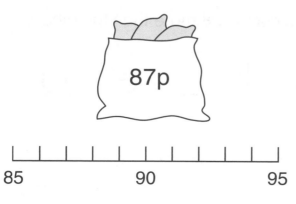

87p   95p

85   90   95

[   ] p

4 Complete this money chain.

£28 → + £5 → [ ]

↳ £40 → [ ] → + £7

↳ [ ] → + £15 ↴

↳ [ ]

Now make up your own money chain.

£ [ ] → + £ [ ] → £ [ ]

↳ + £ [ ] → £ [ ]

↳ + £ [ ] → £ [ ] + £ [ ]

↳ £ [ ]

5 You have £10 to spend.

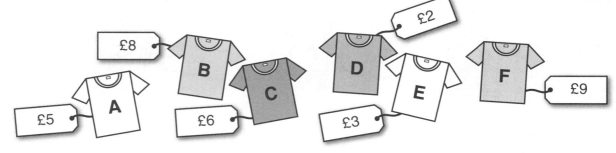

£8 £2 £9 £5 £6 £3

A B C D E F

a Which 2 T-shirts could you buy so that you had £1 change?

[ ]

b Which 3 T-shirts could you buy for exactly £10?

[ ]

1 What is the change from 50p for each amount?
Draw the coins and write the change.

Amount	Coins given as change	Change
38p		
41p		
26p		
29p		
18p		

2 Write the change from £20 for each group of items.

a

| £4 Famous Fives | £8 Detective Stories | £6 Aliens are Coming | £ |

b

| £3 Tumble! | £5 Rainbow DASH | £7 Crunk | £ |

c

| £9 Fluff Series 1 | £2 Double Dunk | £3 Sing Along! | £ |

d

| £5 Boat Jigsaw | £8 Difficult Jigsaw | £4 | £ |

3   Answer these. Complete the bar diagrams to help you.

a   Adam spends a total of 85p on a comic and a drink. The comic costs 50p. What is the cost of the drink?

b   Jade spends a total of 90p on a colouring book and crayons. The colouring book costs 25p. What is the cost of the crayons?

c   Sam spends a total of 99p on a birthday card and stamp. The birthday card costs 49p. What is the cost of the stamp?

d   Mo spends a total of 60p on a pencil case and ruler. The pencil case costs 35p. What is the cost of the ruler?

e   Amy spends a total of 75p on a bottle of water and an apple. The apple costs 15p. What is the cost of the water?

4 Use the items in this sports shop to answer these.

a What is the total cost of buying the net and balls?

£

b How much change from £20 would you get if you bought two bats?

£

c How many bags of balls could you buy with £20?

d What is the total cost of buying the net and table?

£

e Jo bought a net and some new trainers. The total cost was £50. How much did the trainers cost?

£

f 2 tickets, for an adult and a child, to watch a table tennis tournament cost £30. The adult ticket costs £6 more than the child ticket. What is the price of the child ticket?

£